MATH Challenge

Published by InSTEP Education Publishing, Inc.
PO Box 1912 Coppell, TX 75019

ISBN 978-0-9905149-5-4

For information about custom editions, special sales, premium and corporate purchases, please contact InSTEP Education Publishing at (214) 206-5926 or customerservice@instep.education.

MATH Challenge

The exercises within this workbook were created by highly-qualified educators and are designed to provide students practice with a variety of math skills, including fact recognition, basic operations (addition, subtraction, multiplication, and division), pattern recognition, and logical analysis.

The goal of each exercise is to encourage students to think critically to determine patterns and visualize mathematics outside of the standard format. The exercises are divided into three sections based on challenge level, ranging from easy to complex, and increase in difficulty to provide exposure, development, and practice with logical application of mathematics. Encourage your child or student to attempt as many exercises as possible to enhance skills.

Students are also provided the opportunity to apply heavy use of cognitive skills as they create their own puzzles to share with others.

The Math Challenge workbook is great for daily warm-up, enrichment exercises, or regular skill maintenance. We ask the question, ***How many puzzles can you solve?***

Created & Produced by Teachers

InSTEP Education Publishing Inc.
www.InSTEP.Education/publishing

Table of Contents

MATH ★ Challenge

Adding Pyramids

Directions: Can you solve the pyramid puzzles below? Add the endpoints on each side. Write the sums in the middle circles.

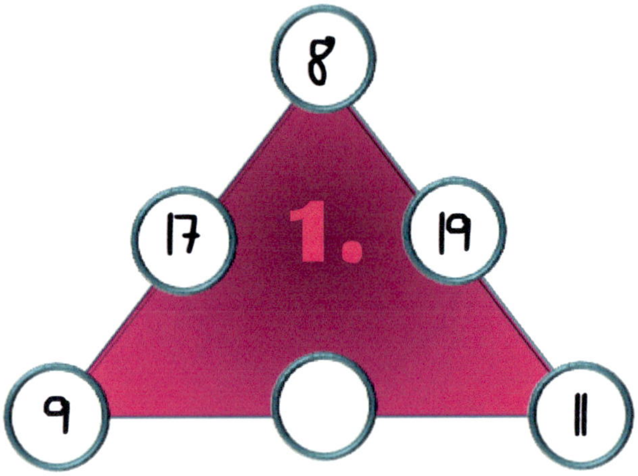

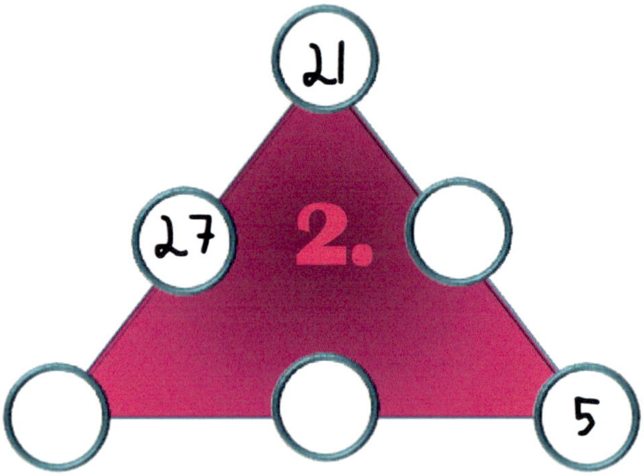

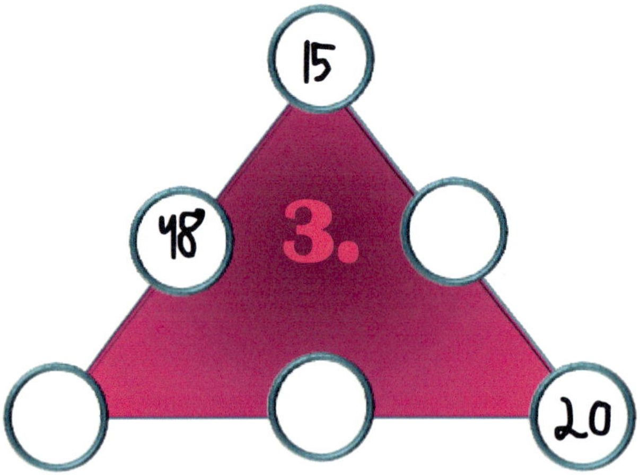

MATH ★ Challenge

A Piece of Pie

Directions: Fill in the missing numbers to complete each pattern.

1.

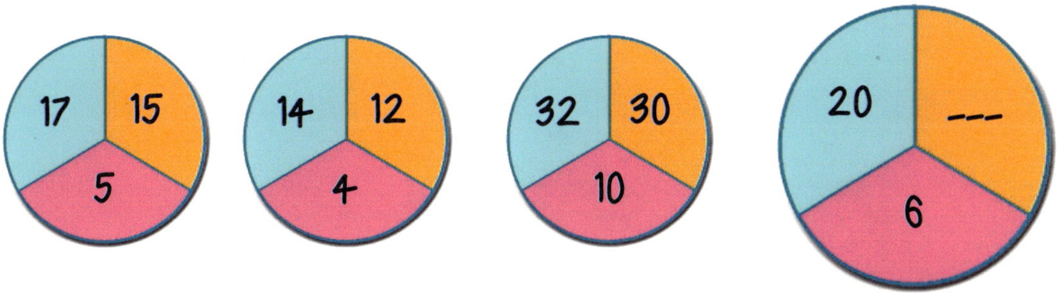

2.

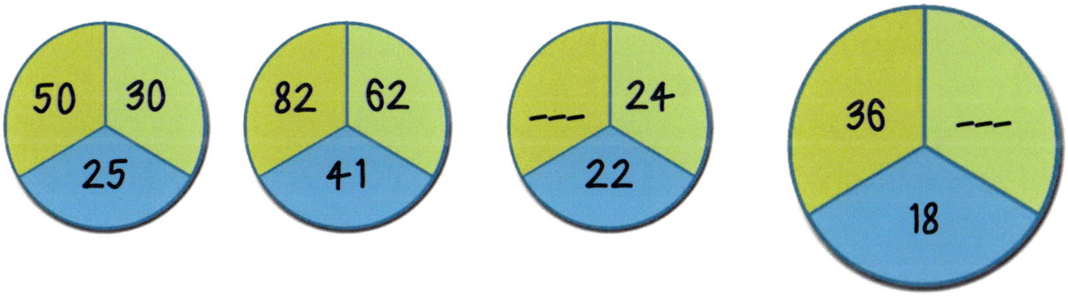

3.

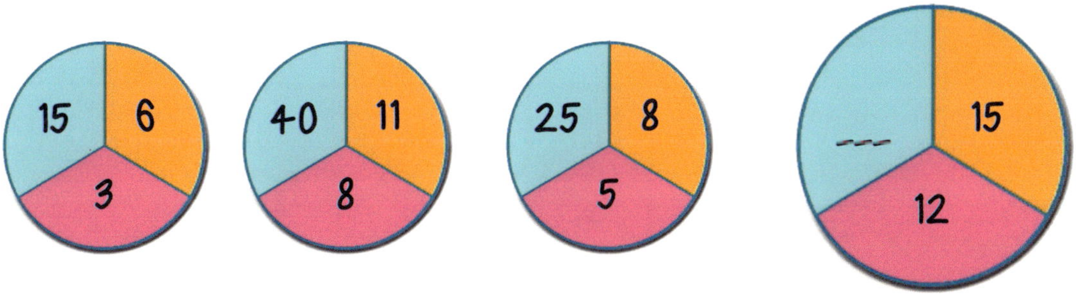

4.

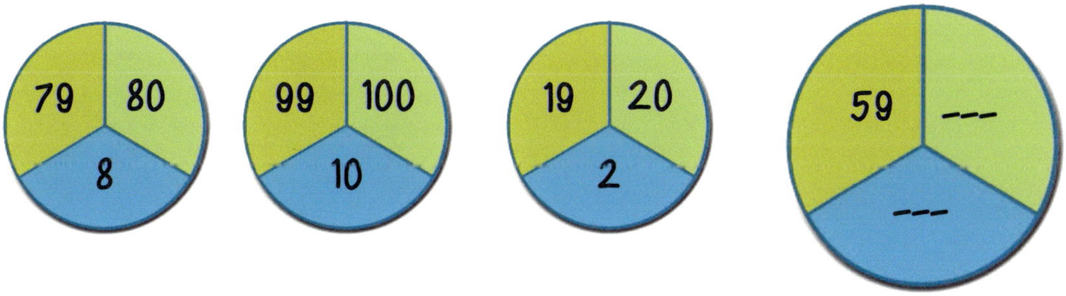

Rows & Columns

Directions: Fill in the missing numbers to make the equation true on each row and column.

	+	7	=	12
+		-		+
11	-		=	
=		=		=
16	+	4	=	

10	+		=	28
-		-		-
	+	5	=	
=		=		=
4	+		=	17

	+	13	=	35
+		-		-
13	-	3	=	
=		=		=
	-		=	

MATH Challenge

Shapes R Us

Directions: Solve each problem below.

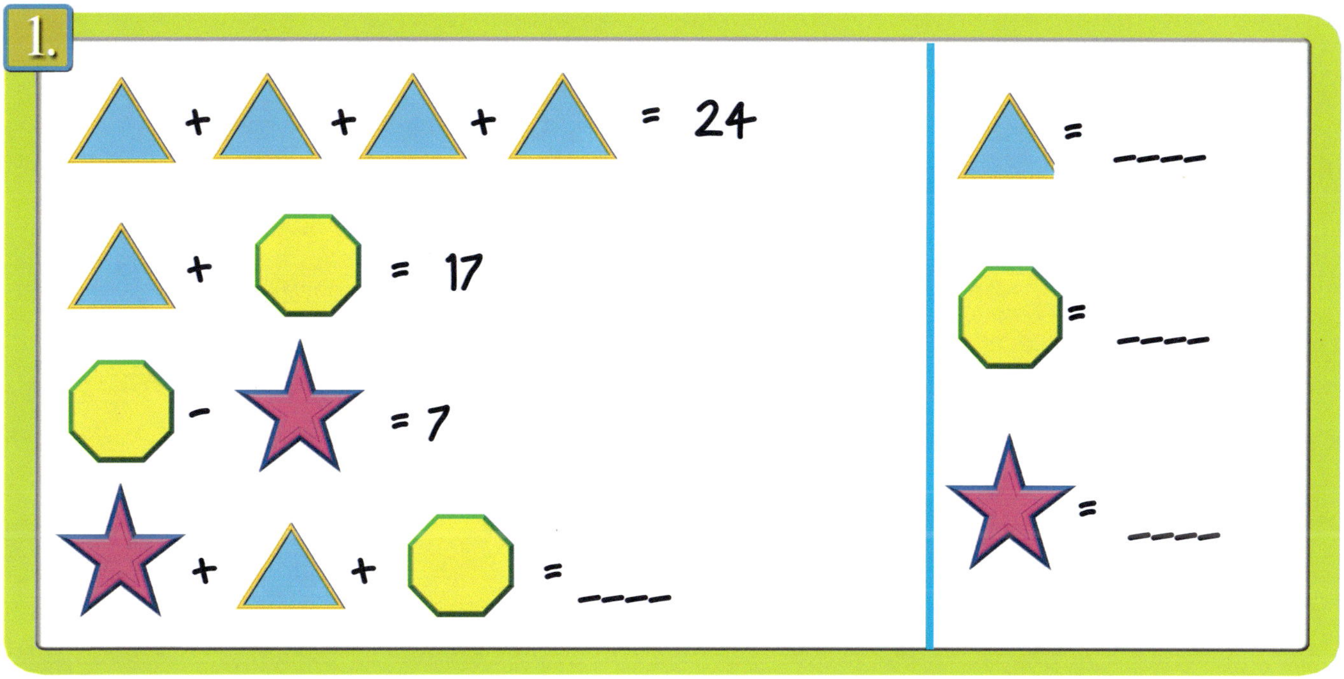

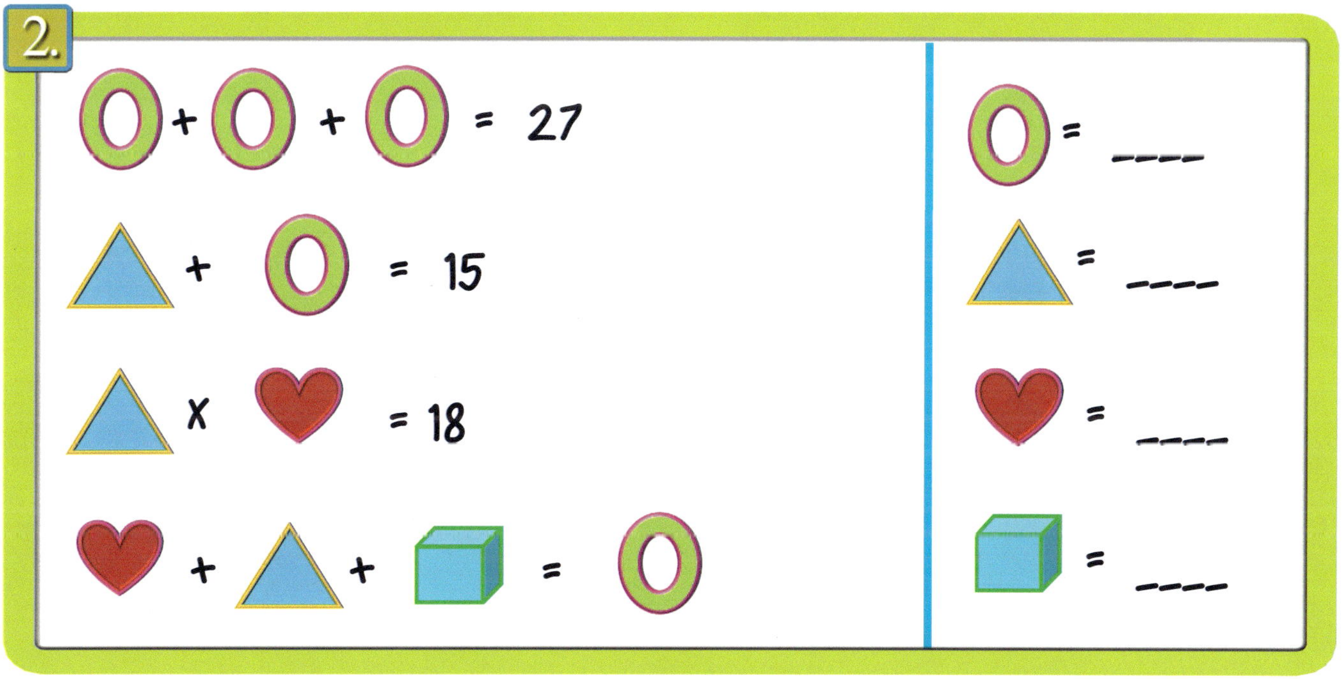

MATH ★ Challenge

Taking Sides

Directions: Add the endpoints to find the sum of each side of the rectangles below. Write the sum of each side in the square boxes.

1.

2.

3.

4.

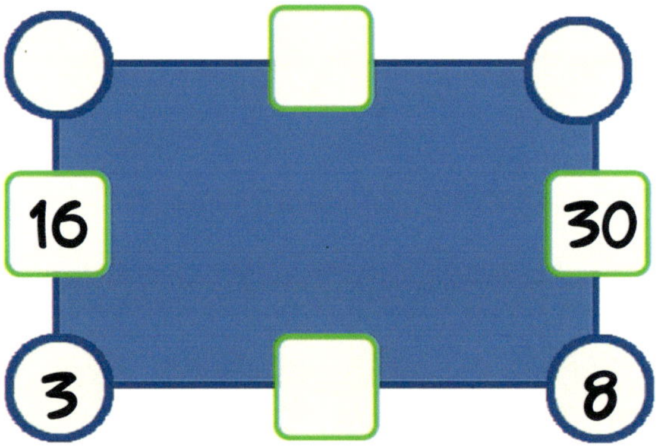

5.

6.

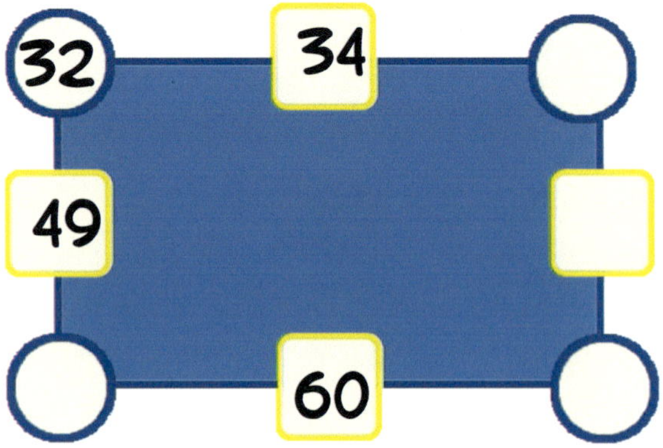

MATH ★ Challenge

Finding Time

Directions: Complete each pattern by writing the correct time on the watches below.

1.

2.

3.

MATH ★ Challenge

All Together

Directions: The sum of each shape below is written in the middle. Can you fill in the missing numbers?

1.

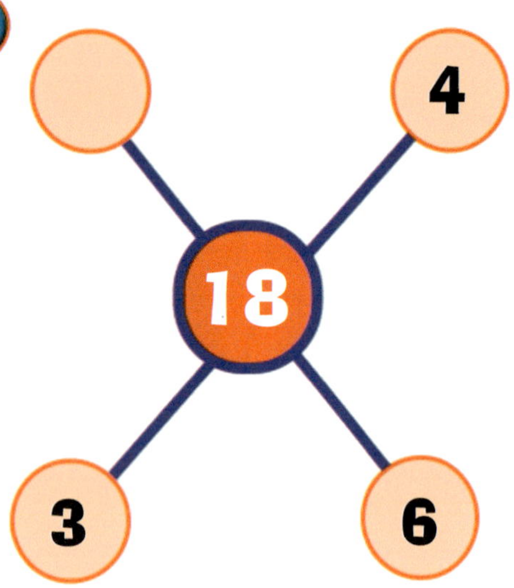

2.

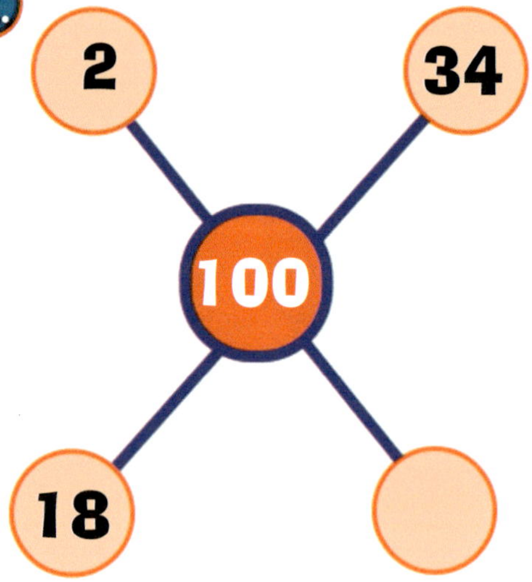

3.

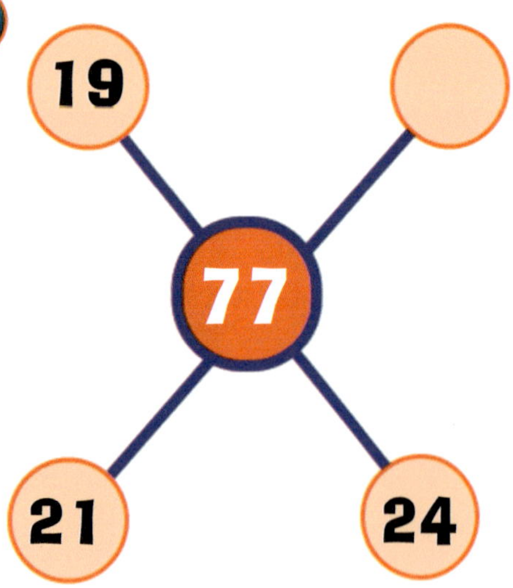

4.

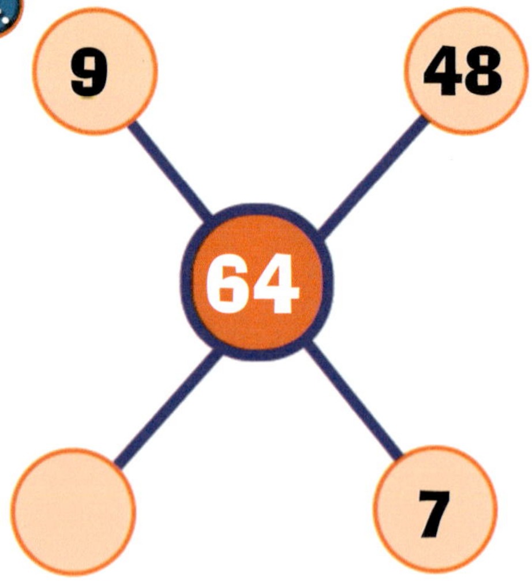

MATH Challenge

Pyramid Puzzles

Directions: Look at the example pyramids to identify the pattern. Fill in the missing numbers below.

1.

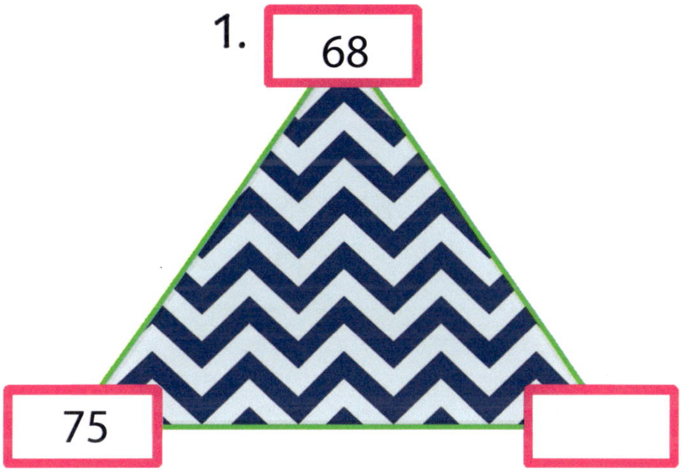

2.

3.

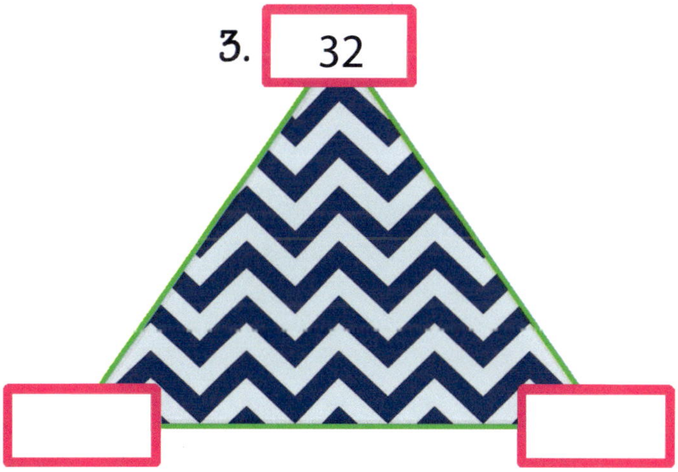

MATH ★ Challenge

Climbing Ladders

Directions: Start from the bottom, and work your way up! Use the given numbers to create an equation that equals 30. You must only use each number once!

1.

1, 2, 5, 10, 60

2.

1, 2, 4, 7, 12

3.

2, 3, 5, 6, 46

Justifying Shapes

Directions: Read each word problem below. Justify your answer in complete sentences.

1. Josh often does yard work for many of the families in his neighborhood. Mrs. Smith's yard is 100 feet wide and 50 feet long. Which of the following figures could be a sketch of Mrs. Smith's yard? ____

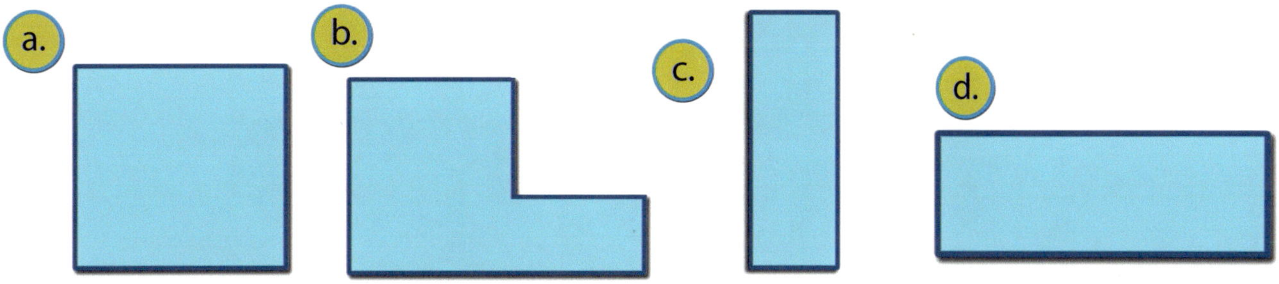

How do you know? Justify your answer here:

____

____

____

2. Josh helped Mr. White build a fence around his vegetable garden. The garden is 25 feet wide and 30 feet long. Using this picture of Mr. White's yard, which section is the vegetable garden?

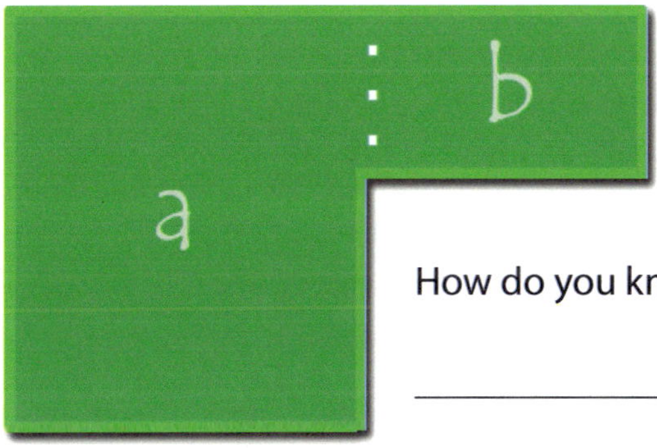

- ○ Section A
- ○ Section B

How do you know? Justify your answer here:

____

____

____

MATH ★ Challenge

Number Chain

Directions: Each new number is created by adding the previous two numbers together. Fill in the missing numbers to complete each chain.

1.	8	9	17				
2.	2	3				21	
3.	6		11	16			70
4.		8	9			43	69
5.	10		30		80		210
6.	7		22		59		
7.	3		79			389	623
8.				60	99	159	258
9.	4	119				892	1,447
10.				90	153	243	

MATH Challenge

Congratulations!!

____
(Name)

has completed and mastered the
Math Challenge Level I!

____________________ ____________________
(Signature) (Date)

Math Riddles

Directions: Can you answer each of the riddles below?

1. What 3 positive numbers give the same result when added together or multiplied?

2. I add 5 to 9 and get 2. The answer is correct, but how?

3. The ages of a father and son add up to 66. The father's age is the son's age reversed.

   How old could they be?

4. If:
   2 = 5,
   3 = 10,
   4 = 17,
   5 = ??

MATH ★ Challenge

Fact Puzzle

Directions: There are 20 addition, subtraction, division, and multiplication facts hidden in the puzzle below. Can you find all of them? Be sure to write each fact you find on the lines.

8	4	32	2	12	8	4	20	13	9
15	12	19	21	6	36	22	18	11	6
5	7	11	7	2	4	49	1	72	54
3	3	2	3	0	9	7	16	7	9
9	4	6	5	30	19	7	13	3	11
10	11	21	1	15	1	36	2	8	8
8	0	6	0	9	3	27	6	4	1
6	11	13	8	2	8	20	8	2	7
4	18	1	18	13	6	10	23	6	15
24	3	18	6	3	48	2	9	4	5

1.______________________________

2.______________________________

3.______________________________

4.______________________________

5.______________________________

6.______________________________

7.______________________________

8.______________________________

9.______________________________

10.______________________________

11.______________________________

12.______________________________

13.______________________________

14.______________________________

15.______________________________

16.______________________________

17.______________________________

18.______________________________

19.______________________________

20.______________________________

MATH ★ Challenge

Robot Mania

Directions: Can you find the total area and perimeter of the robot below?

Total Area: ________

Perimeter: ________

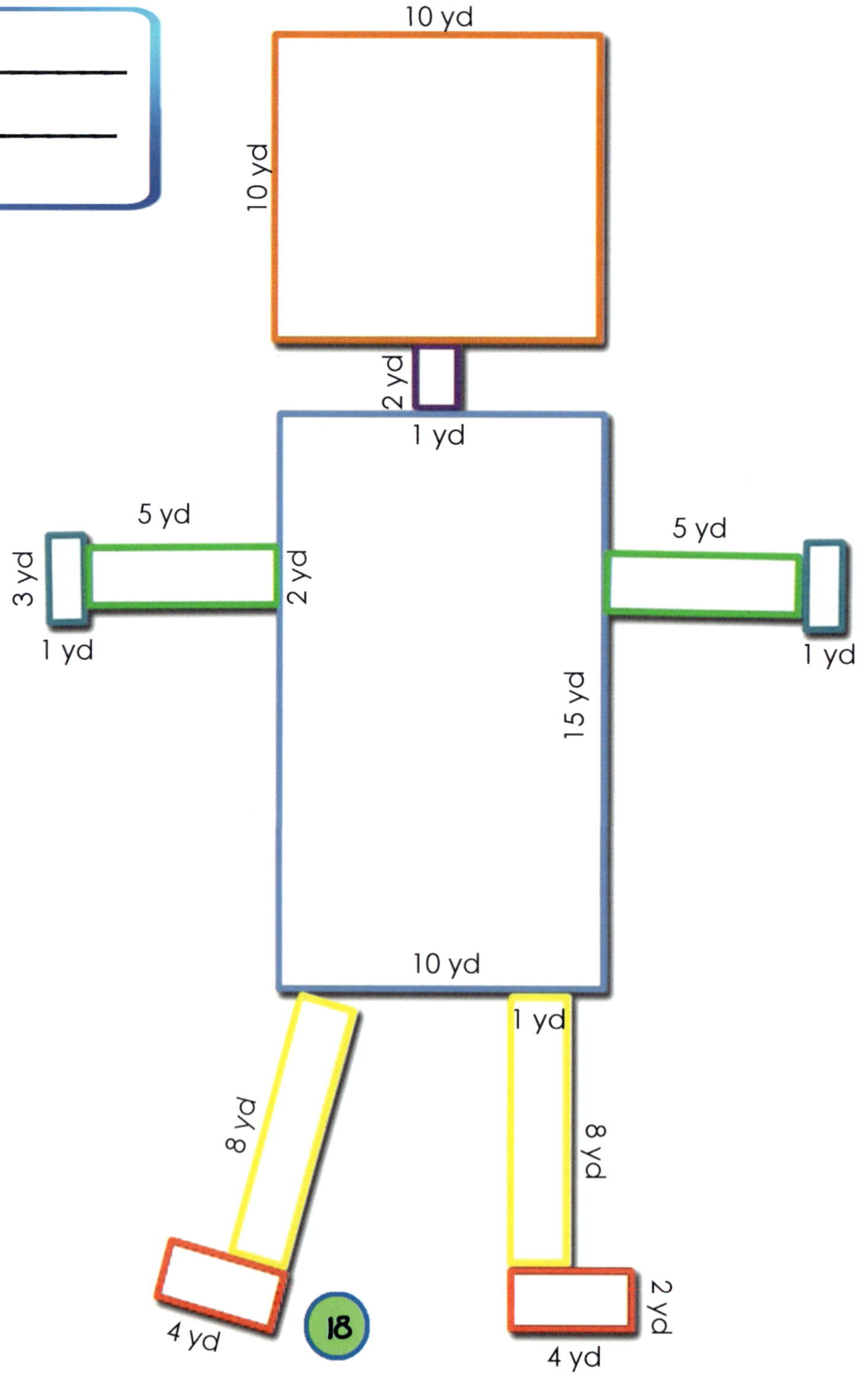

MATH Challenge

Sports Sum It

Directions: Complete the chart below by filling in each row and column with the missing sums.

 = ____________

= ____________

 = ____________

 = ____________

Shapes 4 U

Math Challenge

Directions: Solve each problem below.

1.

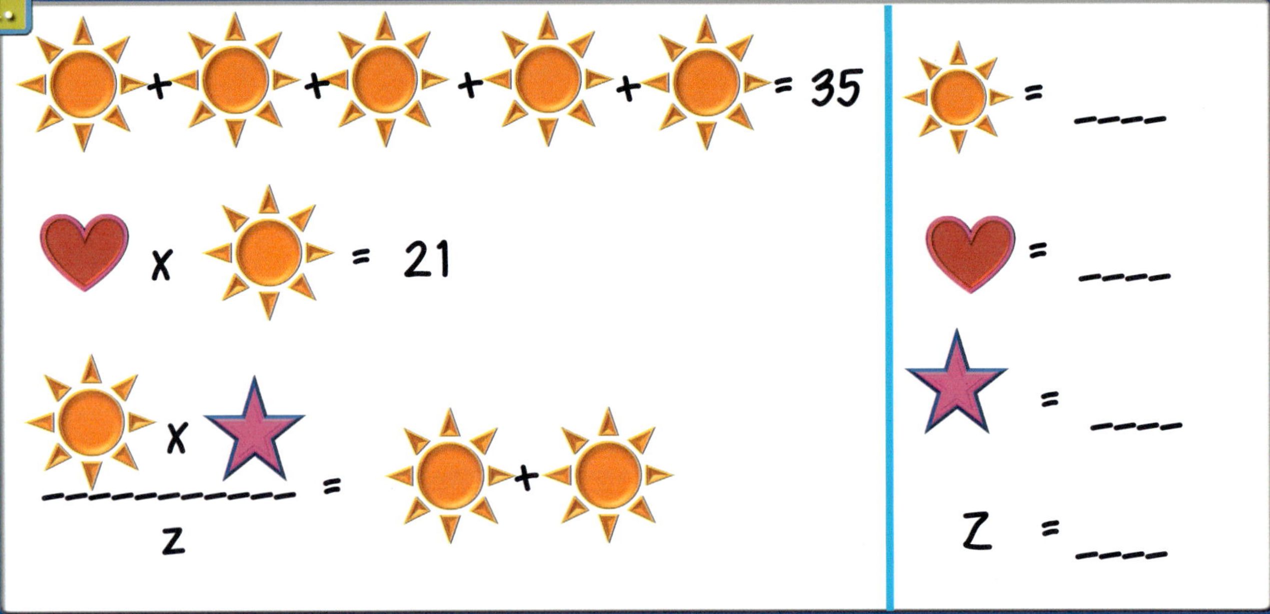

2.

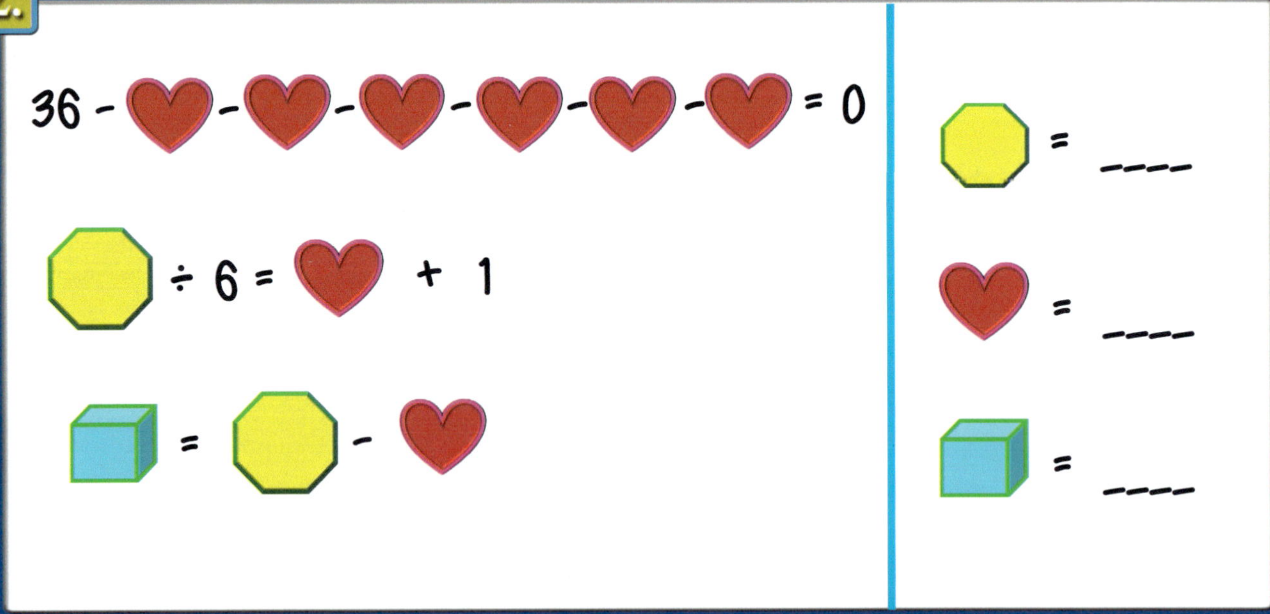

MATH ★ Challenge

Pyramid Puzzle 2

Directions: Look at the example pyramids to identify the pattern. Fill in the missing numbers below.

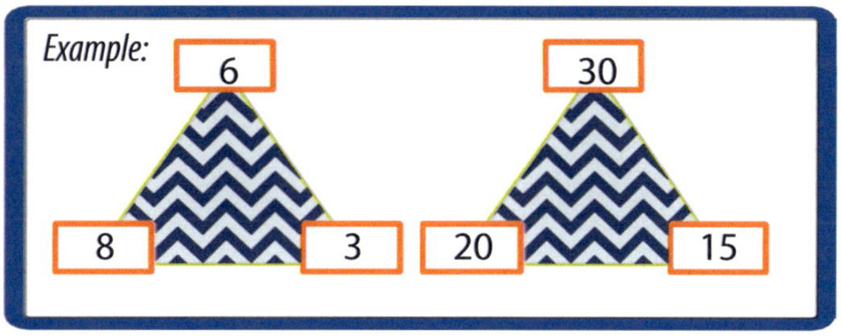

1.

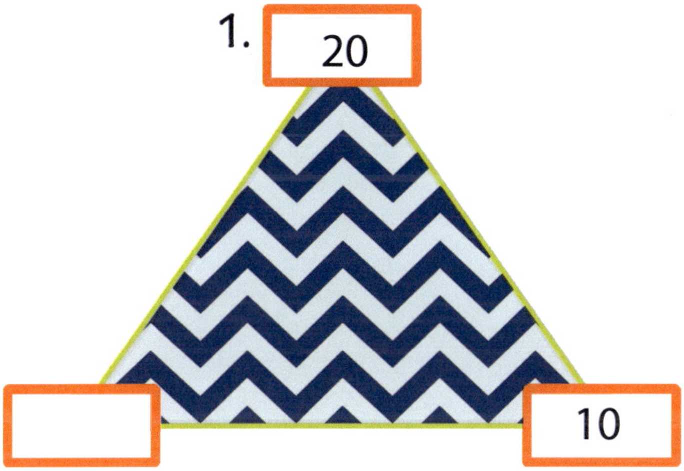

2.

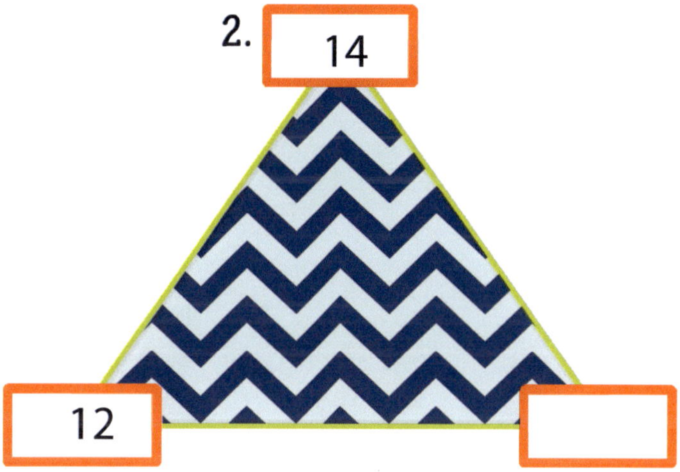

3.

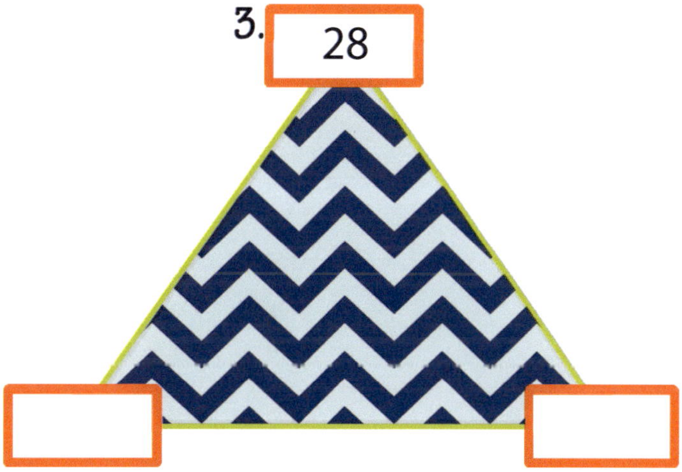

Factor Trees

Directions: Can you complete the factor trees? Use the numbers in the box to complete each factor tree below. The first one has been completed for you.

MATH ★ Challenge

Sparkly Sums

Directions: Complete the chart below by filling in each row and column with the missing sums.

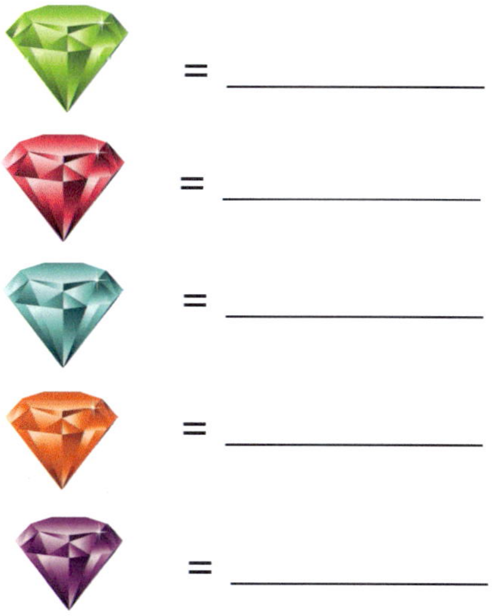

Area & Perimeter

Cutting Up is a popular lawn care service used by many homes and businesses. They charge $2 per foot of yard each month to cut and maintain the lawns of their customers. Usually they charge three times the perimeter of the lawn to plant hedges and bushes around the lawn.

Directions: Read each word problem. Use the information above to answer the questions.

1. The lawn in front of Mr. Pickney's office is square and measures 8 feet on each side. How much does Mr. Pickney pay each month to maintain his office lawn?

2. The figure below is a sketch of the Tolbert family's backyard.

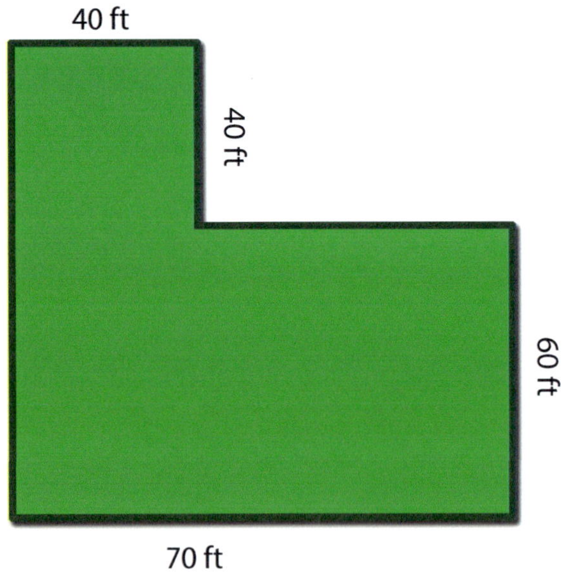

a. How much will the Tolbert family pay to have Cutting Up plant hedges around their entire yard?

b. What is the total area of the Tolbert family's yard?

MATH Challenge

Congratulations!!

____
(Name)

has completed and mastered the
Math Challenge Level II!

____
(Signature)

____
(Date)

MATH ★ Challenge

Number Sentences

Directions: Can you make number sentences using the cars below? Write +, -, ÷ , or = the boxes below to make each number sentence true. You may use operations more than once, if necessary.

1.	20	☐	4	☐	1	☐	6
2.	30	☐	2	☐	5	☐	3
3.	5	☐	1	☐	4	☐	16
4.	3	☐	21	☐	6	☐	5
5.	63	☐	9	☐	42	☐	6
6.	6	☐	19	☐	5	☐	4
7.	121	☐	0	☐	113	☐	8
8.	48	☐	4	☐	12	☐	144

MATH Challenge

Pyramid Puzzles 3

Directions: Fill in the missing numbers in the pyramids below.

1.

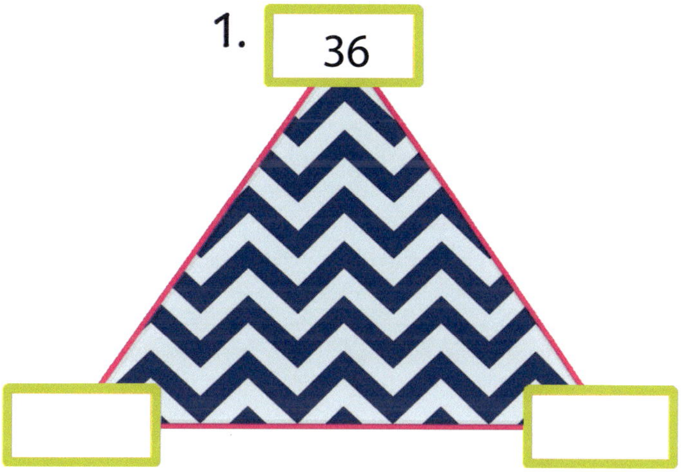

2.

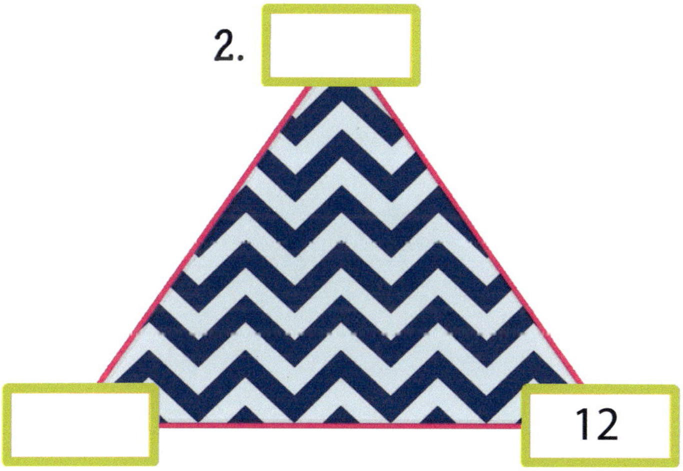

3.

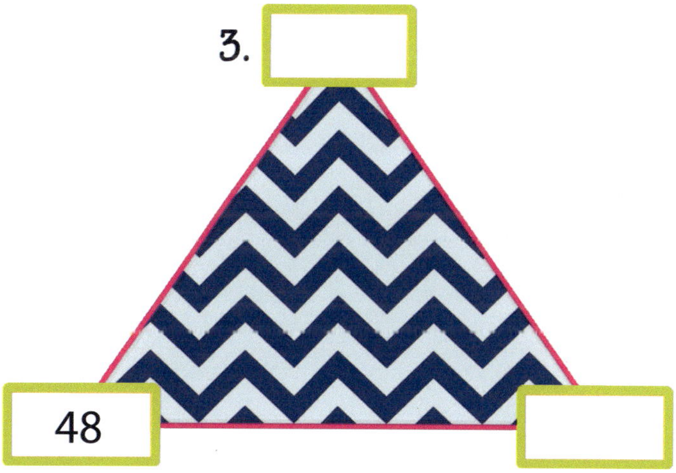

MATH ★ Challenge

Sum in the Middle

Directions: Using the numbers 1-10, fill in the empty spaces so the sum of each row and column equals the number in the middle. You may use each number only once.

1.

	12	

2.

	13	

3.

	14	

4.

	15	

MATH ★ Challenge

Step Up To 50

Directions: Start from the bottom, and work your way up! Use the given numbers to create an equation that equals 50. You must only use each number and operation once!

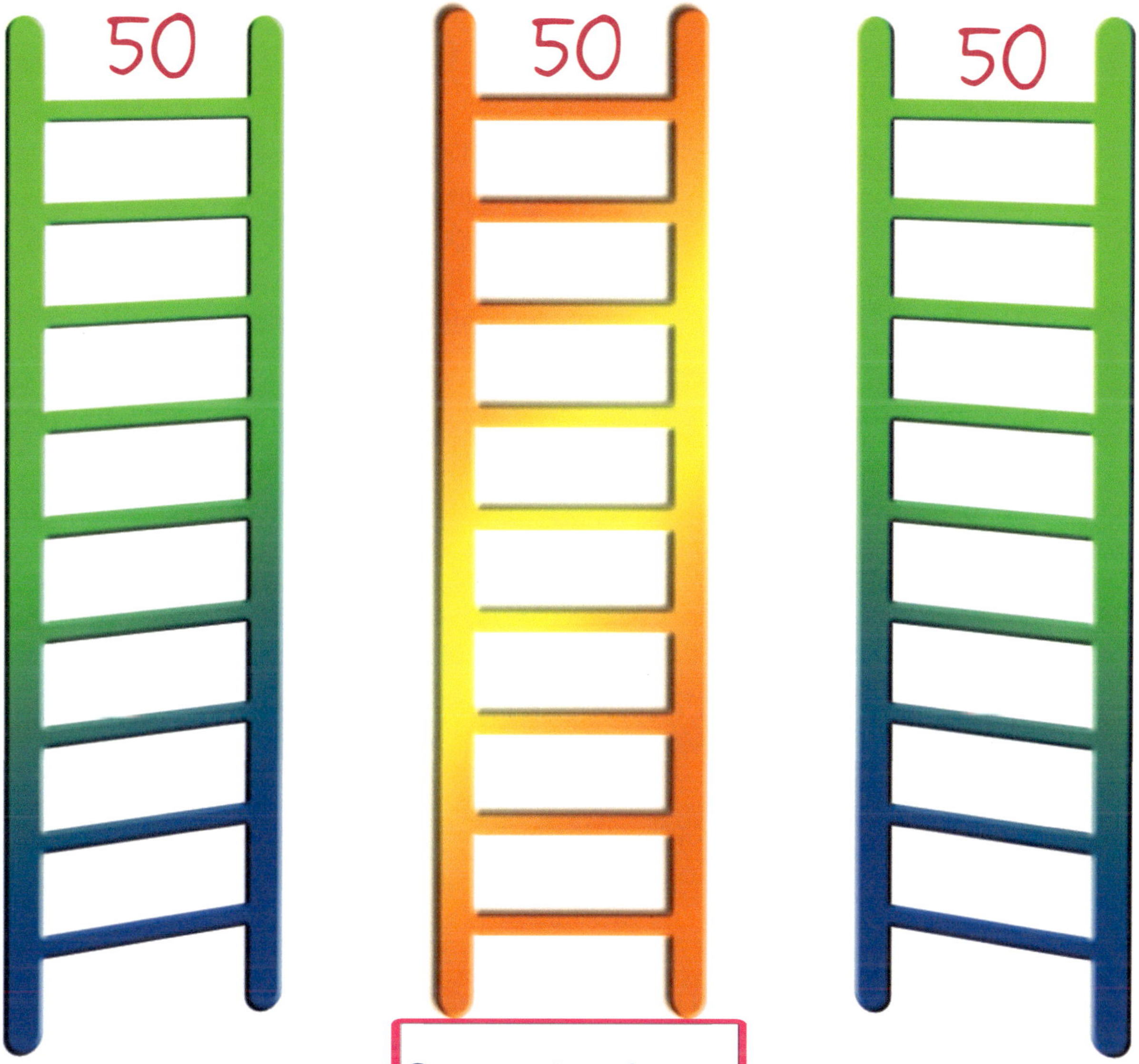

1. 1, 5, 5, 10, 11

2. 2, 5, 6, 10, 45

3. 1, 5, 8, 15, 25

Sum Around The Box

Directions: Follow the directions given in the rectangles below.

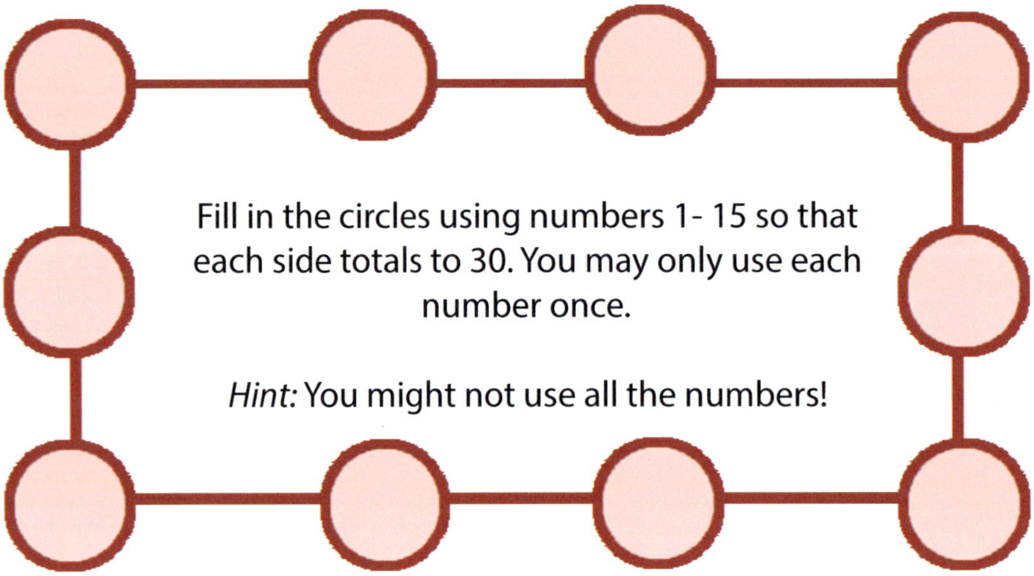

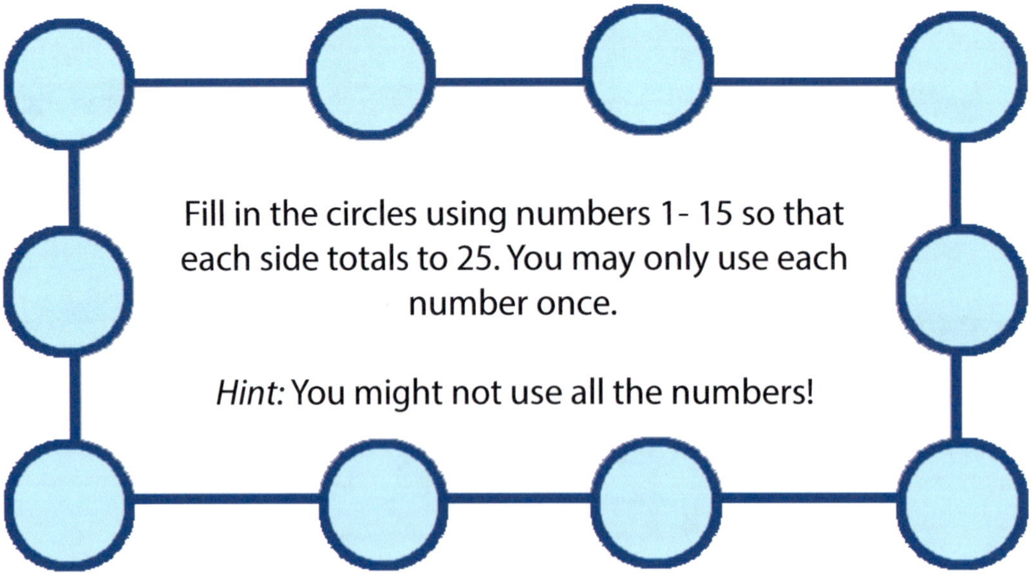

Area & Perimeter 2

PJ and Alisha Kidd are the owners of "Kidding Me", a small store that sells clothes and shoes for kids in their area. They want to make changes in and around the store. Use the floor plan to answer the questions below.

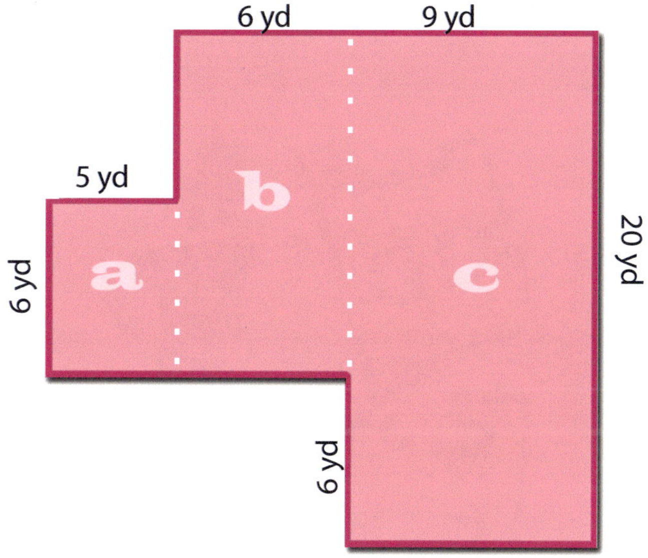

1. The Kidds would like to put wood floors in area C of the store. How much wood flooring will they need to purchase?

____

2. The Kidds would also like to paint a really cool border on the walls of areas A and B. If she painted one long continuous stripe, how long would that border be?

____

3. Alisha Kidd is considering painting the ceilings in areas A & C. If a gallon of paint covers 100 yd^2, how many gallons will she need to buy?

____

4. PJ Kidd likes the idea of putting a sidewalk outside around the store, so he can exercise daily by walking. What distance would he walk if he walked 5 times a week?

____

5. The owners may be interested in buying a larger space. What is the total area of Kidding Me's current location?

____

MATH Challenge

Wow, Super Congratulations!!

____
(Name)

has completed and mastered the
Math Challenge Level III!

____________________ ____________________
(Signature) (Date)

MATH ★ Challenge

Create Your Own

A Piece of Pie

Directions: Devise your own rules to create the puzzle below. Challenge a friend to solve it!

1.

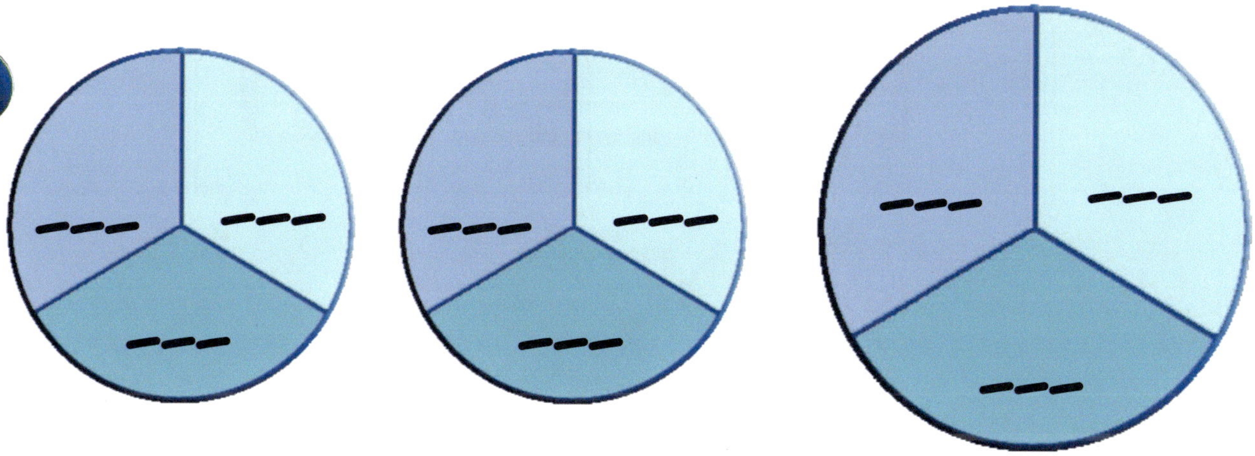

2.

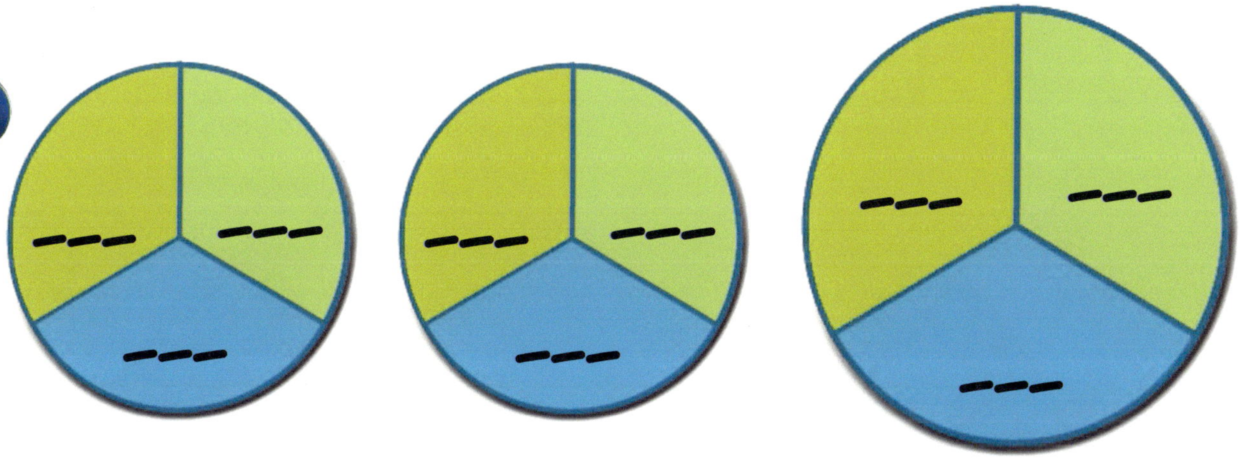

MATH ★ Challenge

Create Your Own Puzzle Pyramid

Directions: Devise your own rules to create the puzzle below. Challenge a friend to solve it!

Example:

1.

2.

3.

Create Your Own

Number Sentences

Can you make number sentences using the cars below? Write +, -, ÷ , or = the boxes below to make each number sentence true. You may use operations more than once, if necessary.

Directions: Plan the numbers needed to create the puzzle below. Challenge a friend to solve it!

1. 

2. 

3. 

4. 

5. 

MATH ★ Challenge

Create Your Own Fact Puzzle

Directions: Hide ten addition, subtraction, multiplication, or division facts create your own fact puzzle below. Challenge a friend to locate all ten facts to solve it!

1.____________________

2.____________________

3.____________________

4.____________________

5.____________________

6.____________________

7.____________________

8.____________________

9.____________________

10.____________________

pg. 4- Pyramid Puzzles

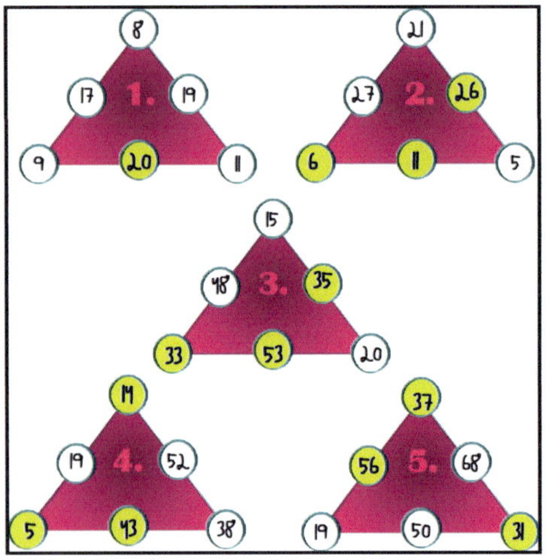

pg. 5- A Piece of Pie

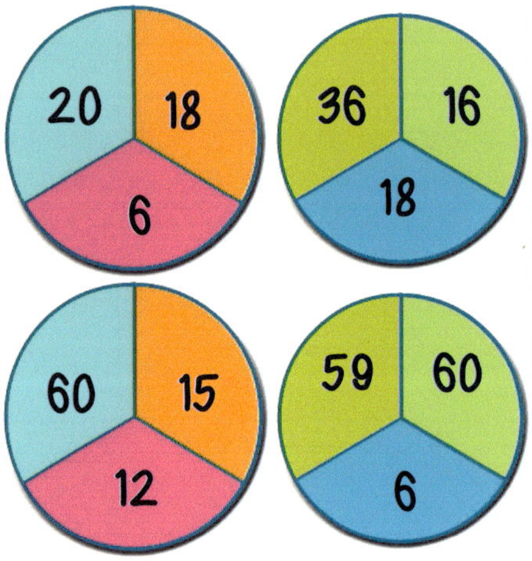

pg. 6- Rows and Columns

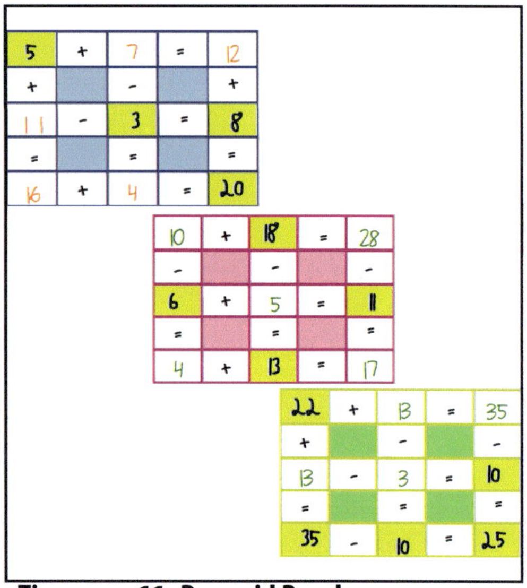

pg. 7- Shapes R Us

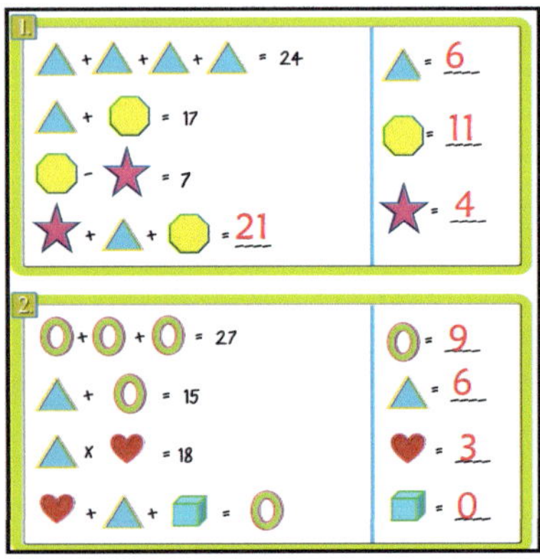

pg. 8- Taking Sides

pg. 9- Finding Time

1. 7:45
2. 10:20
3. 3:05

pg. 10- Sum It Up

1. 5
2. 46
3. 13
4. 0

pg. 11- Pyramid Puzzles

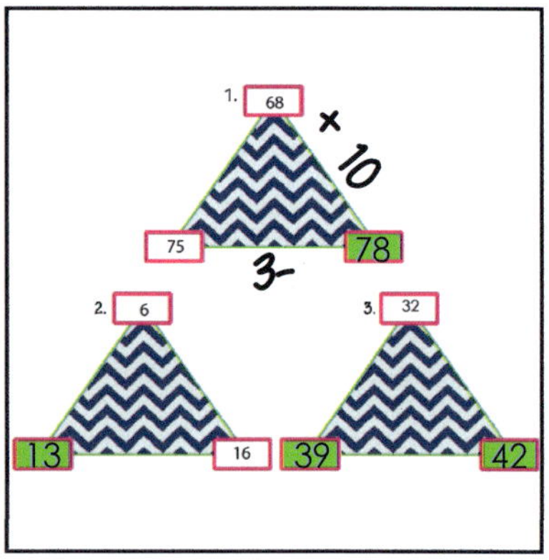

pg. 12- Climbing Ladders

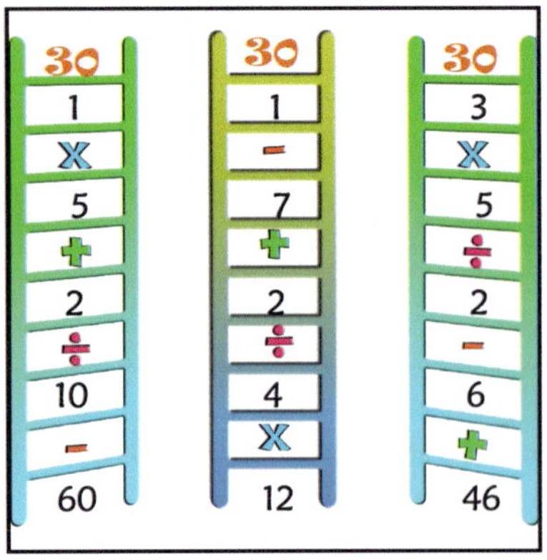

pg. 13- Justifying Shapes

1.d - because the shape described is a rectangle, with a greater width. Shape a is square (have the same width and length), shape b has too many different lengths, and shape c has a greater length than the width.

2. a- because the shape described is nearly equal in length and width (almost square). Section b has a width much greater than the length.

pg. 14- Number Chain

1.	8	9	17	26	43	69	112
2.	2	3	5	8	13	21	34
3.	6	5	11	16	27	43	70
4.	1	8	9	17	26	43	69
5.	10	20	30	50	80	130	210
6.	7	15	22	37	59	96	155
7.	3	76	79	155	234	389	623
8.	18	21	39	60	99	159	258
9.	4	119	218	337	555	892	1,447
10.	36	27	63	90	153	243	396

pg. 16- Math Riddles

1. 1, 2, 3

2. You must think about time on a clock. 5 hours after 9 am is 2 pm.

3. 42 and 24

4. 2, because 2 = 5.

MATH Challenge

Answer Key

pg. 17- Fact Puzzle

8	4	32	2	12	8	4	20	13	9
15	12	19	21	6	36	22	18	11	6
5	7	11	7	2	4	49	1	72	54
3	3	2	3	0	9	7	16	7	9
9	4	6	5	30	19	7	13	3	11
10	11	21	1	15	1	36	2	8	8
8	0	6	0	9	3	27	6	4	1
6	11	13	8	2	8	20	8	2	7
4	18	1	18	13	6	10	23	6	15
24	3	18	6	3	48	2	9	4	5

pg. 18- Robot Mania

Total Area = 310 yd^2

Perimeter = 212 yd

pg. 19- Sports Sum It

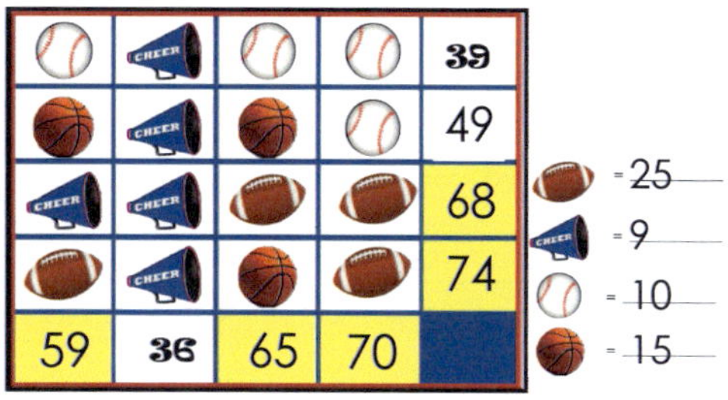

pg. 20- Shapes 4 U

pg. 21- Pyramid Puzzles 2

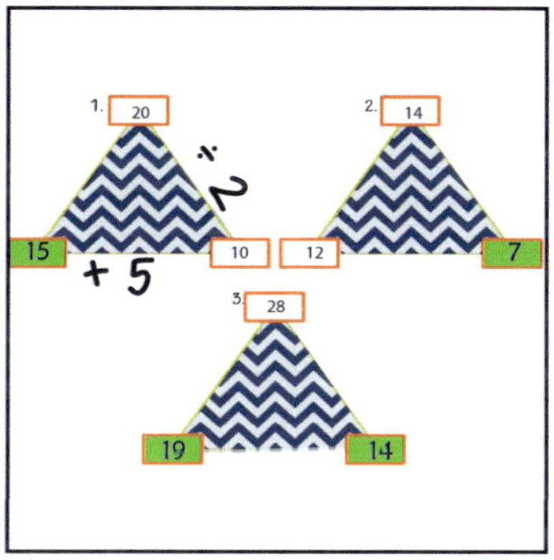

pg. 22- Factor Trees

pg. 23- Sparkly Sums

pg. 24- Area & Perimeter

1. $128

2 (a). $1,020

(b). 5,800 ft^2

pg. 26- Number Sentences

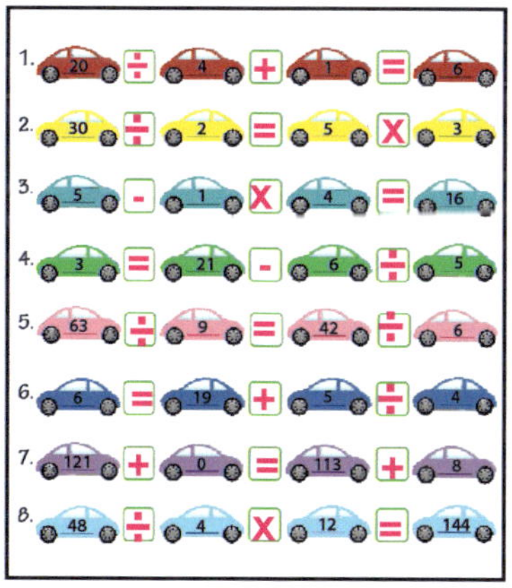

MATH Challenge

Answer Key

pg. 27- Pyramid Puzzles 3

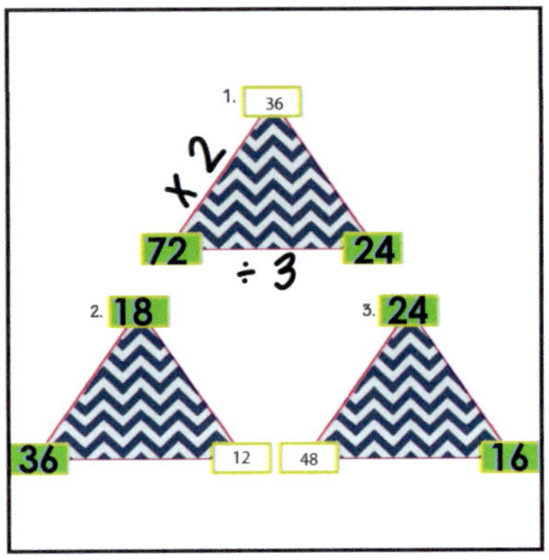

pg. 28- Sum in the Middle

possible answers:

1.

3	7	2
8	**12**	4
1	5	6

2.

7	4	2
5	**13**	8
1	9	3

3.

7	5	2
6	**14**	8
1	9	4

4.

2	7	6
10	**15**	5
3	8	4

pg. 29- Step Up to 50

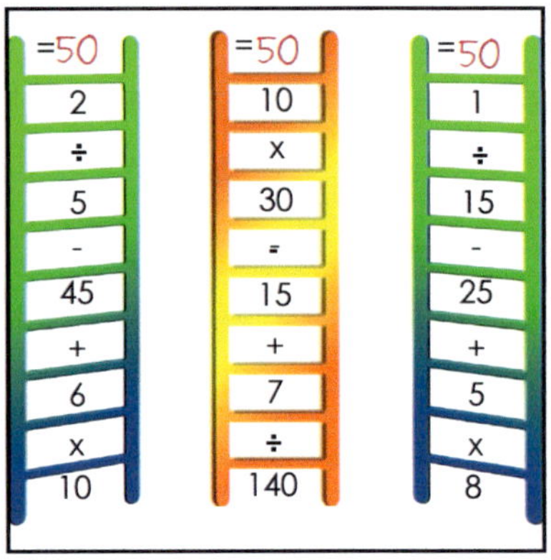

pg. 30- Sum Around the Box

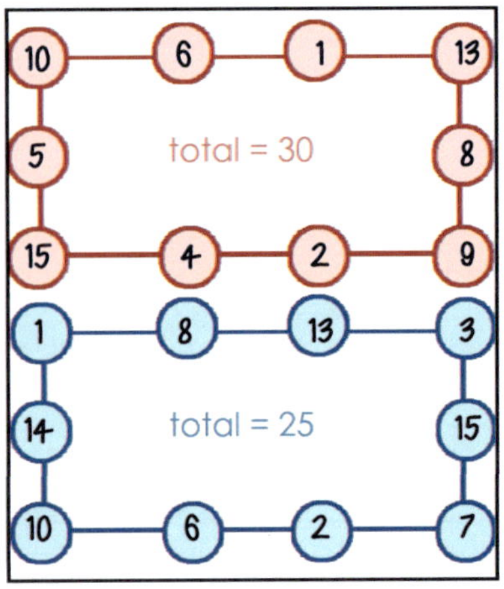

pg. 31- Area & Perimeter 2

1. 180 yd^2
2. 36 yd
3. 3 gallons (area of A & C = 210 yd^2)
4. 400 yd (perimeter = 80 yd)
5. 294 yd^2

Made in the USA
Monee, IL
05 November 2026

67377440R00026